Bibliographic information published by the German National Library:

The German National Library lists this publication in the National Bibliography; detailed bibliographic data are available on the Internet at http://dnb.dnb.de .

Imprint:

Print and binding: Books on Demand GmbH, Norderstedt Germany
ISBN: 9783346046161

This book at GRIN:

https://www.grin.com/document/503718

Musaad Alruwaili

From Tanker Trucks to Trains. Safety in Oil Transportation

GRIN Verlag

Table of Contents

Executive Summary

The overall objective of this report was to identify an existing safety problem in the company's oil transportation operations and provide recommendations on the strategy the company could implement to plug the identified weakness. Effective, efficient, and safe transportation of oil is not only one of the most important aspects of the company's long-term growth prospects, but it is also one of the integral aspects of the development of the environment and, by extension, the economy and the society. Despite the company's awareness of the need for effective, efficient, and safe strategy for oil transportation, it is heavily reliant on a mode of oil transportation (tanker trucks) that it is anything, but safe, efficient, and effective.

Based on these drawbacks, this report recommends that the company switch from tanker truck transportation to rail transport. In contrast to tanker truck transportation, transportation of oil through rail is effective, efficient, and safe. Studies indicate that rail transport is a safer mode of transport because trains transporting oil have a small risk of colliding with other trains when transporting oil to their intended destination. The reduced risk of collision means that their impact of trains on the environment from oil spillage is low. In addition to safety, rail transport is a more effective and efficient mode of transporting oil because of the absence of obstacles on the rail truck that might severely interrupt the transportation of oil. However, the oil corporation must proceed with caution during the switch because it will need $15 billion to make the move successful.

Summary of Problem and Recommendations

- **Purpose**

The overarching aim of this report is to identify inefficiencies that the oil company is facing in its crude oil transportation operations and recommend a solution that will assist it in addressing those weaknesses. The company is heavily reliant on tanker trucks, which according to recent studies are an inefficient, ineffective, and hazardous technique of crude oil transportation. The studies indicate that tanker trucks are perilous to the environment because they have a high risk of road collisions (Crawley, 2016; Inderwildi & King, 2012). Some studies indicate that a large percentage of road collisions involving tanker trucks have resulted in the spillage of oil on roads, drainage areas, and other parts of the surrounding environment (Bai & Bai, 2014). Other studies suggest that the road collisions can lead to an inferno that visits disastrous consequences on the lives of motorists and pedestrians. Additionally, tanker trucks are inefficient and ineffective because heavy traffic and the bulky nature of crude oil interrupts the speedy transportation of crude oil to its intended destination. This report will underline the types of weaknesses attendant to tanker truck transportation and make the case for a switch to rail transportation. It report will present evidence to back its assertion that railway transport is the most efficient, safe, and cost-effective technique of crude oil transportation.

- **Literature Review**

Although the crude oil industry has many options when it comes to crude oil transportation, tanker trucks are the most popular technique for transporting crude oil. In tanker truck transportation, the company does a large percentage of its crude oil transportation on five axel trailers with tank-shaped carriers (Nasir, 2015). The companies utilizing tanker trucks for their transportation needs assert that they prefer it because they are flexible in the way in which they aid them in establishing a seamless connection between the pipelines in the oil extraction site and the oil refineries (Nasir, 2015). They assert that the seamless nature of that extraction-refinery connection arises from the manner in which the tanker trucks move to and from the stationary pipelines without necessitating significant interruption in the process of transferring oil from the extraction site to the transportation mode (Nasir, 2015). This argument suggests that tanker trucks do not require significant interruption in the flow of oil into the oil tanks. As long as the company has positioned many trucks on close to the crude oil pipelines, the flow of oil into the tanks used to transport oil to the refineries will be smooth. Further, the argument suggests that other modes of transport would require significant interruption during the process of transferring crude oil from pipelines into the designated modes of transport. In fact, the companies confirm this when they assert that although individual tanker trucks carry a small load, they prefer them because of the efficient way in which they permit the movement of crude oil from the pipelines to the oil tanks.

Despite these claims on the advantages of using tanker trucks, studies on their drawbacks raise doubts about their suitability for the transportation of crude oil from the extraction sites to the refineries. The studies suggest that tanker trucks are laden with perils that make their use as a mode for crude oil transportation hazardous, inefficient, and ineffective. In particular, the studies indicate that tanker trucks are extremely hazardous. In particular, Fannelop (2013) states that the risk of a BLEVE (Boling-Liquid-Expanding-Vapor-Explosion) is higher at the loading and offloading terminals because the extraction and transportation processes contribute to the heating of the crude oil above its boiling point. He argues that their explosion risk at the designated regions for loading and discharging crude oil is extremely high (Fannelop, 2013).

Other studies (Crawley, 2016; Inderwildi & King, 2012) indicate that poor management of the loading and discharge terminals can lead to the exposure of the crude oil to fire sparks, which can occasion an explosion. The studies argue that this high risk of explosion at the loading and discharge terminals is one of the factors that continue to undermine the suitability of tanker trucks as a crude oil transportation mode.

In addition to the hazards lurking at the loading and discharge terminal, studies (Tamminen, 2012; Renne, 2013) have used inefficiencies at the loading and discharge terminals to question the suitability of tanker trucks in crude oil transportation. (Tamminen, 2012; Renne, 2013) argue that weaknesses in the design of the pipes used to load and offload crude oil into and out of the tanks leads to a high degree of inefficiency in process of transporting crude oil from the pipelines at the extraction site to the pipelines at the refinery. (Tamminen, 2012; Renne, 2013) argue that weaknesses in the design of those pipes leads to the spillage of large amounts of crude oil and delays in transferring crude oil into the tanks. Oil companies lose million from the resultant spillage and delays in moving oil to the refineries. This inefficiency has forced many scientists to argue that flaws in the design of the infrastructure used to load and offload crude oil into and out of the tanks make tanker trucks an inefficient means of crude oil transportation.

Additionally, the high spillage risk and its adverse effect of tanker truck related spill has influenced experts (Tamminen, 2012; Fannelop, 2013) to question their suitability as an effective mode of transporting crude oil to refineries. Tamminen (2012) is one of the experts who have used the high oil spillage risk of tanker trucks to question their effectiveness as a mode of transporting crude oil. Tamminen (2012) contends that a single tanker truck accident can contribute to the spillage of a significant volume of crude oil into waterways. To underscore the high spillage risk, Tamminen (2012) highlights a six-month period in 2000 when three tanker truck accidents on Californian roads contributed to the spillage of 12,850 gallons of oil into nearby waterways. This outcome highlights how the high spillage risk can occasion adverse effects on the environment. In this case, three tanker truck accidents were enough to contribute to the spillage of more than 12,000 gallons into waterways in California. The elevated spillage risk makes tanker trucks an ineffective approach for the transportation of crude oil.

Further, the (Renne, 2013; Tamminen, 2012; Morris, 2012) have identified high accident risk in their analysis of the factors that continue to undermine the appropriateness of using tanker trucks in crude oil transportation. The studies (Renne, 2013; Tamminen, 2012; Morris, 2012) suggest that the risk of accidental collisions between crude oil tankers and other vehicles. The studies (Renne, 2013; Tamminen, 2012; Morris, 2012) indicate that the risk of accidental collisions reaches its peak during the high season when the crude oil companies ask the drivers to drive their trucks to and from the refineries at a faster speed to satisfy demand for petroleum products. This leads to a spike in the incidence of head-on collisions between tanker trucks and other vehicles. Morris (2012) confirmed the link between careless driving and tanker-truck head-on collisions when he pointed out that the collisions increase the risk of tanker truck explosions. When sparks from the collision strike the oil spilling from the tanker truck, the result is usually an explosion that places the tank driver, pedestrians, and other road users at an elevated risk of death. This high collision risk has influenced scientists to argue that tanker trucks are a hazardous mode of crude oil transportation.

Apart from tanker trucks, experts (Molinski, 2015; Bai & Bai, 2014) in crude oil transportation have raised doubts about the suitability of the use of pipelines in crude oil transportation. The experts argue that while pipelines are an efficient method of transporting crude oil, they are fraught with several safety risks that make them an inappropriate transportation mode. Molinski (2015) contends that the oil leakage risk in pipeline transport is higher than the leakage risk in other modes of transport because of the difficulties oil companies face in cleaning and servicing their oil pipelines. Further, the leakage risk is high because companies face challenges in identifying leaks in their pipeline (Molinski, 2015). Bai and Bai (2014) argue that the high leakage risk in pipeline transportation arises from natural disasters, external loads, third-party damage, external corrosion, or internal corrosion. The subsurface leakage of oil can cause a significant problem to the oil company because it can cause delays in oil production, contaminate the underground water table, and threaten ecosystems (Bai & Bai, 2014). In fact, the companies can take up to two months to identify and correct leaks in their pipelines. In many cases, the companies identify the risk after a substantial volume of the oil has leaked into the surrounding environment and the underground water table and contaminated the soil and water that the neighboring community depends on for their nutritional needs (Bai & Bai, 2014). Such a substantial leak will be costly to the affected company because it will force it to use a significant proportion of its income in cleaning up the leak and settling claims relating to the leak. The high risk of oil leaks and the subsequent cost of repairing the damage caused by the leak suggest that pipeline transport joins tanker trucks in the list of ineffective and hazardous methods of crude oil transportation.

- **Proposed Solution**

The review of literature has offered insight into the factors that have influenced crude transportation experts to conclude that tanker trucks and pipelines are inefficient, ineffective, and hazardous methods of transporting crude oil. This section of the report proposes the solution for transporting crude oil in an effective, efficient, and safe manner. To be more specific, the section gives reasons to substantiate the claim that rail transport is an efficient, effective, and safe method of transporting crude oil.

Low spillage risk is one of the factors that crude oil transportation experts (Crawley, 2016; Wardley-Smith, 2012; Molinski, 2015) highlight when presenting arguments about the good safety record railway transport. The experts assert that the tankers used in railway transport are tougher and heavier than the tanks utilized by tanker trucks. According to Crawley (2016), stakeholders in the oil and gas sector have moved to reduce the risk of spillage in rail transportation by focusing their efforts on derailment prevention and tank design. The focus on tank design has contributed to the manufacture of tanks that have thermal protection, pressure relief valves, full-height head shields, and outer steel jackets (Crawley, 2016). The focus on derailments has led to the development of trains with superior braking power, wheel-bearing detectors, and effective traffic routing sensors (Crawley, 2016). These innovations have not only reduced the derailment risk in railroad transport, but also significantly reduced the risk of oil spillage and tank explosion during derailment. In fact, the innovations around tank design have reduced the risk of tank explosion in situations where the tanks are in high temperature environments or environments with sparks and other chemical compounds that may ignite the crude oil. The toughness of the material makes it harder for tanks to spill their contents during train derailments (Molinski, 2015). In fact, the number of tanks spilling oil in a single train derailment is usually smaller and the volume of oil spilled into the environment during a leak is often lower than the volume of crude oil spilled after tanker truck accidents. The experts argue that the low spillage risk makes railroad transport the least hazardous approach for transporting crude oil from the extraction site to the refinery.

The reduced spillage risk is also advantageous to the company from the perspective of reduction in wastage. The durability of the tanks used in the transportation of oil on train carriages reduces the losses that the company would face because of lost supplies (Wardley-Smith, 2012). Oil companies that rely on tanker trucks experience heavy losses because of spillage and tanker truck explosions (Wardley-Smith, 2012). The cost of delays, lost supplies, and replacing tanker trucks can undermine their ability to generate above average returns from the sale of crude oil (Wardley-Smith, 2012). Similarly, oil companies that rely on pipeline face a significant risk of financial loss because of rising compensation claims (Wardley-Smith, 2012). In contrast, companies transporting oil through the railroad do not incur similar costs during a derailment because of low spillage risk. This low risk implies that the company will only incur costs associated with delays in delivering the oil to the refinery (Wardley-Smith, 2012). The reduced wastage on railroad transport makes this method of ferrying crude oil superior.

Apart from low spillage risk, oil and gas experts (Crawley, 2016; Renne, 2013) contend that railroad transport is effective because of its ability to transport large volumes of crude oil in a single trip. Unlike tanker trucks, train carriages can carry a larger volume of oil in one trip. Esser (2014) contends that the volume of crude oil transported by rail transport per trip is so significant that it is comparable to oil transported by way of pipelines. Indeed, Crawley (2016) offers support to the arguments about the capacity of railroad transportation of crude oil in his arguments about the trends in fossil fuel extraction. According to Crawley (2016), many oil companies in the East Coast, West Coast, and Gulf of Mexico coast have abandoned pipelines and moved to railroad transportation because of its effectiveness in transporting large volumes of crude oil. Crawley (2016) contends that railroad transport is so effective that it assists oil companies to transport 450 million barrels of crude oil. Renee (2013) substantiates this contention by stating that the double and triple tracking of 18,000 miles of railroads in the US has increased railroad capacity, increased crude oil transportation speed, and lowered oil companies' operating costs. These arguments suggest that, when properly utilized, railroad transport can offer the same efficiency benefits as pipeline transport, but with a spillage risk that is superior to that of tanker trucks.

The limited impact of rail transport on the environment during the process of ferrying crude oil offers further insight into its suitability as a transportation mode. One of the issues that proponents of railroad transportation have raised during their arguments on the suitability of railcars in crude oil transportation is their minimal impact on the environment (Inderwildi & King, 2012). Railcars do not produce the same volume of pollutants during crude oil transportation as other transportation modes (Inderwildi & King, 2012). The movement of 1 ton of crude oil from the extraction site to a refinery several miles away requires less energy and has lower emissions than alternatives like tanker truck transportation (Inderwildi & King, 2012). This advantage means that railcars are effective and efficient because they permit the transportation of oil at the lowest cost to the environment (Inderwildi & King, 2012). The social benefit of transporting crude oil through railcars means provides additional incentive for companies to switch from tanker trucks to railcars. Such a switch will offer benefits to the environment and improve the reputation of oil companies at a time when many international organizations are associating them with climate change and global warming.

- **Conclusion**

The overall aim of this report was to propose a mode of crude oil transportation that was effective, efficient, and safer than tanker truck transportation. This objective arose from the company's complaints about its tanker trucks safety record. An analysis of studies on suitability of tanker trucks as a mode of crude oil transportation has demonstrated that it is highly inefficient, ineffective, and hazardous. The studies indicate that tanker trucks have an elevated risk of explosion, a high collision risk, and are hazardous to the environment. Switching to railroad transportation would offer the company significant efficiency, effectiveness, and safety benefits. Efficiency and effectiveness benefits would arise from the way in which reliance on railcars would permit the company to transport larger volumes per trip. Safety arises from the innovations in tanker design and the efforts to reduce the derailment risk of railcars.

- **Recommendation**

Tanker trucks are not the only means of transporting oil. Apart from tankers, one can use the train or water vessels. Unlike pipelines and tanker trucks, rail and water transportations means are more flexible and cheaper. According to Tippee (2012), the use of rail transport reduces shipping costs when compared to pipeline transport. In addition to being practical, rail and water transportation are cost-effective. Nonetheless, it is important to note that each means of transportation differently affects public health, the environment, and safety. In this perspective, a risk is a possible occurrence of an incident. Of the three methods of transporting oil, the company should choose rail transport to carry its oil. Busteed (2016) asserts that the transportation of oil by rail has become an increasingly common method and is presently responsible for the transportation of more than 10% of all oil in the U.S. This shows that railroads possess some advantages over other modes of transporting crude oil. The next paragraphs offer an in-depth explanation of railroad.

The company should use rail transport because of its several benefits. Rail transport is cheap: the company will only use 1-2 million dollars per every mile. Special trains will also be used to transport oil. Several firms involved in oil transportation have adopted rail transport. According to Busteed (2016), there has been a significant increase in the utilization of railroad in the transportation of oil. Firms have chosen railroad transportation because of its versatility and its safety record. Besides, the percentage of oil spills on railroads is lower than that experienced on motorways and pipelines. However, railroad transportation has some disadvantages. Examples include train derailment, poorly designed tank cars, and badly assembled trains: a major cause of train derailment . These threats disproportionately affect human beings. According to Liu, Saat, and Barkan (2017) train derailment damages the infrastructure, harm the environment, and cause casualties. While railroad transportation has a number of threats, they can be solved.

Pipelines utilized to transport crude oil require regulatory oversight (Busteed, 2016). This calls for high quality planning from local authorities. For example, pipelines should not be placed close to residential houses or undertake their activities closer to the railroads because of the risk of explosion. In a company, the health of the public should always be the first priority. By safeguarding the life and property of the people, the company will be safeguarding its public image.

Mitigation to disasters occasioned by railroad is a form of emergency preparedness. Each stakeholder in the oil transportation should always be prepared for emergency cases. According to Busteed (2016), an emergency response team should be given weekly estimates of the number of trains carrying crude oil as a way of preparing them to respond to an oil spill. A full study on the perils of railroad transportation of should be undertaken beforehand to develop superior strategies that can prevent the issues once they happen. Equally, mechanisms that can help to effectively handle any mishaps should be formed. Comprehending the nature of the crude oil can also help to avert cases of explosions on railroads. Since the routes and schedules of trains are not disclosed beforehand as a safety measure against terrorism, the alertness of the emergency response team may be hampered. To circumvent this problem, a select team of responders should be conscious of the train routes and schedules beforehand. Technology can also be used to curb incidents of train derailment. For example, a mobile applications can provide emergency responders with information regarding train incidents. Such mobile apps offer emergency responders with the opportunity to quickly know about a train accident and offer quick rescue to those affected by the incident. With such technology, the security of rail transport is enhanced. However, no safety measures have been implemented in the case of truck transportation of crude oil. This implies that the train is the better option of transporting oil. Besides, the transportation of oil by train should be governed by new rules.

Introducing new rules to govern the utilization of trains in transporting oil will increase the safety of railroad. There is an ongoing debate regarding rules to guide the transportation of oil through trains. The new guidelines stipulate that new brake should be installed in trains as a way of minimizing the dangers caused by highly flammable hazardous materials (Busteed, 2016). However, there are agitations to change the new guidelines. Regardless of the challenges, the safety administration is against changing these rules since their objective is to decrease incidents likely to occur because of transporting oil through trains. According to Busteed (2016), opposition towards new rules has increased doubt regarding the transportation of oil via trains. Once the company has sanctioned and adopted the use of railroad, new strict guidelines should be formed before the use of trains.

- **Leading Safety Indicators :**

The practice that involves planning rather than wait for bad things to occur may be applied to the workplace safety, using leading indicators. Leading indicators refer to pre-incident measurements, which are collected after the occurrence of an incident and are in contrast to lagging indicators (Morrison, 2014). Michelle Garner-Janna, Cummins Inc.'s director of corporate health and safety, points that when one waits for the occurrence of an incident before they can make a change, it may be too late. While researchers tend to disagree on what constitutes leading indicators, leading indicators are considered as measurements of safety events and behaviors that precede occurrences and have a predictive feature (Morrison, 2014). Measuring the probability of occurrence of unsafe things and taking necessary steps to prevent them can assist safety professionals to keep employees safe in the workplace. This paper discusses safety-leading indicators of railroad transportation, which include near misses, hazards, maintenance, training, and management of change.

- **Near Misses**

Reporting near misses is an important way of identifying weaknesses in the rail transport worksite. Near misses can be described as events that fail to meet the organization's criteria for recordable events like oil spills (International Safety Association of Oil and Gas Producers 2011). Near misses are also able to provide observations of unsafe conditions, which may not have consequences. Such conditions are categorized as events that could have potential, in other different circumstances, to lead to results that might have been recordable, especially high-potential events, in which significant incidences are real worst-case scenarios (International Safety Association of Oil and Gas Producers 2011). As such, near miss events can give leading information regarding the probability of actual incidents as well as offer lagging information about barrier weaknesses. In addition, near miss examination significantly contributes to the continuous improvement of process safety and asset integrity, if it is used in identifying weakness and as warning of potential catastrophes.

- **Maintenance**

Maintenance indicators compare actual conditions with specific targets or requirements. Maintenance indicators consist of realistic goals and baselines to facilitate diagnostic and prognostic processes and justify related decisions and actions at appropriate organization's levels (Liyanage & Kumar, 2003). Maintenance indicators are associated with the improvement of health, safety, productivity, quality, and capacity utilization, as well as the reduction of costs, wastes, and downtime. Examples of maintenance indicators include budget maintenance, possible targets, and the period between breakdowns and repairs (Liyanage & Kumar, 2003). Today, most senior management in the rail transport seeks to understand the value that is created by maintenance processes, while taking into consideration safety issues.

- **Training**

Training employees to conduct themselves in a safe way and use all equipment appropriately is considered a leading indicator since it correlates to the safety of facilities. When more employees are trained, the safer the work environment will be (Morrison, 2014). A statistical analysis conducted at Cummins found training hours to be a leading indicator and had the strongest relationship to lagging indicators (Morrison, 2014). Where more hours were used to conduct training, fewer occurrences of safety incidents were recorded. Training as a leading indicator also entails more than just signing workers up for classes. It is also important to monitor training programs to ensure that they are conducted effectively and everyone attends (Morrison, 2014). This will result in improved safety at the workplace people will have more aware of dangers and hazards. However, training does not simply end with monitoring attendance or the training sessions. The training can be useful if it is thoroughly conducted and if the management observes that people are getting involved in unsafe practices, then there is some ineffectiveness in training.

- **Management of change**

Rail transport organizations have safety responsibilities and have to consider and manage any possible risks to commuters, employees and the public from new changes. Change that may include risk to these groups are likely to arise from organizational, technical (engineering), and operational changes (Liyanage & Kumar, 2003). As such, a critical addition to the regulatory regime can be introducing *Common Safety Method* on evaluation and analysis of risk.

Appendix

- **Background Information**

The inspiration to write this proposal was provided by the necessity of the company to enhance the safety of transporting oil. Previously, the company relied on tanker trucks to transport its oil. However, tanker trucks could not satisfy company's safety requirements. Explosions and fires of tanker trucks made the company incur immense losses. Equally, several people (company employees and members of the public) died because of the explosion of tanker trucks. Besides, oil spillage from the tanker trucks is a hazard to the environment. Apart from being a danger to the environment, oil spillage is a loss to the firm. To avert further injury to employees and accidents, the company should enact this new proposal. Safety, for both employees and the public, should be the number one priority for any company. Failure to implement the proposed change will imply that the firm will continue experiencing deaths through their carelessness. In the long term, the needless deaths occasioned by accidents of tanker trunks will lead to a total collapse of the company.

- **Literature**

Unlike other methods of transporting oil, trains are faster. It can take many days to move oil through coastline as opposed to the use of pipelines. It is important to note that oil is not a perishable good that will require high speed. The speed of the rail is high considering the fact it is only transporting oil, a non-perishable good. For the safety of the firm, the train is the best plan. Transporting oil at a moderate speed will also ensure that it reaches the offloading point at the exact time. Alternatively, the speed at which the oil is transported should not be perceived as a critical component in oil firms. Rather, the safe delivery of petroleum from one side to the other should assume precedence in all companies. Prioritizing the safe delivery of oil will guarantee that the safety of the company is always maintained at all times: the major aim of undertaking this proposal.

The use of trains in the transportation of oil has also non-price advantages. According to Busteed (2016), increased flexibility is a major non-price benefit of railroad transportation when compared to its alternatives. Owing to their flexibility, trains can quickly react to any market changes. Besides, railroads have shorter commitment contract periods than pipelines and tanker trucks (Gallamore & Meyer, 2014). This makes trains flexible regarding the services provided by the diverse rails. As opposed to the use of tanker trucks, the firm can transit from receiving services from one train firm to the other. If the company is unsatisfied with one Train Company, the short contracts will make it easy to change to the other. All the company needs to do to change to a new train provide is fail to renew the old contract. Since rail companies will always wish for their contracts to be renewed, they will strive to provide the best possible services. According to Gallamore and Meyer (2014), the flexibility of the minimal transit times between the train's destination and the point of departure demonstrates the flexibility of trains. The duration taken to move oil from one place to another is shorter than that used in transporting the product using alternative modes of transportation.

Giving more people control of the train can also increase the safety of railroad transportation of oil. This can help avert the high number of dangers involved in the use of trains. According to Dorrian, Hussey, & Dawson (2007), fatigue is a major cause of train derailment. As opposed to one person, many individuals should accompany the driver as a way of reducing accidents occasioned by attention lapse. Besides, the technical team should always be available to validate that the technical components of the train are operational. For example, a large percentage of railroad accidents that entail freight carriages are caused by the application of the handbrake. Equally, a routine inspection of the train before it departs from the station can significantly help to reduce instances of train mishaps (Busteed, 2016). The inspection should also be undertaken once the train returns from its destination. Once the technical crew guarantees that the brakes of the train are function before it departs from the station, a large percentage of mishaps that occur because of technical errors can be averted.

- **Requirements**

The successful implementation of this proposal depends on the successful implementation of several aspects. The company needs to invest $15 billion into the completion of railroads covering approximately 1000 miles because each mile costs roughly $1-2 million dollars. Equally, $20,000 is needed to finalize this research proposal. The achieve the major goal of this research project, I will also have to support it with the past performance of the firm and that of similar corporation that utilize railroads. Adhering to the recommendation will be extremely helpful in promoting the company's safety. However, the company's executives have the leeway to pick the best mode of transporting oil. Besides, the construction of the stations where the trains will be uploading and offloading oil is crucial in the successful implementation of this proposal. To fulfill this requirement, a large piece of land will be required. Employees will also be trained on how to manage the new methodological details of using railroads.

- **Option Comparison**

When tanker trucks and railroads (the two alternative forms of oil transportation), railroad is the safest. While the proposal promotes the adoption of railroads as opposed to tanker trucks, it still has several disadvantages as mentioned earlier. However, the proposal has explained several strategies of negating the disadvantages of railroad transportation of oil. The presence of disadvantages implies that the company will still encounter problems once it starts using the new solution. These issues will be minimized if the firm is keen on using railroads to transport its oil. Besides, the government should also diligently undertake its oversight role as a way of reducing the limits of railroad transport. Regardless of the issues of this mode, railroads are a short-term solution that should be substituted once a safer means of transporting oil has been identified.

- **Costs**

A majority of major changes in a company are accompanied by high costs. Consequently, moving from mode of transporting oil to the other will be costly to the company. Equally, the company will have to dispose the tanker trucks previously utilized in the transportation of oil. However, each disposal of the tankers will be a loss to the firm. Moreover, the faults in a number of tankers imply that the company will sell them at a reduced price. Selling such assets at a cheaper price will reduce the profits of the company. Conservative estimates show that the firm may only earn $2500 from the sale of the tankers. Conversely, implementing railroad transportation is extremely costly. Rough estimates predict that the company may spend roughly $15 billion to ensure that the rails are operational. Regardless of the costs implications, a comparison of railroad and tankers shows that railroads are ultimately cost-effective. Buying tracks is more expensive than implementing the proposed plan of using railroads in the transportation of oil.

References

Bai, Y., & Bai, Q. (2014). *Subsea pipeline integrity and risk management.* New York: Elsevier.

Busteed, E. (2016). Bakken crude and the Ford Pinto of railcars: The growing need for adequate regulation of the transportation of crude oil by rail. Villanova Environmental Law Journal, 27(1), 63-100.

Crawley, G. (2016). *Fossil fuels: Current status and future directions.* Hackensack, NJ: World Scientific.

Dorrian, J., Hussey, F., & Dawson, D. (2007). Train driving efficiency and safety: Examining the cost of fatigue. *Journal of Sleep Research, 16*(1), 1-11.

Esser, C. (2014). Rail vs. pipelines: how to move oil. *International Energy Agency* Retrieved from http://www.iea.org/ieaenergy/issue6/rail-vs-pipelines-how-to-move-oil.html

Fannelop, T. (2013). *Fluid mechanics for industrial safety and environmental protection.* New York: Elsevier.

Gallamore, R., & Meyer, J. (2014). *American railroads: Decline and Renaissance in the twentieth century.* Massachusetts: Harvard University Press.

Inderwildi, O., & King, D. (2012). *Energy, transport, and the environment: Addressing the sustainable mobility paradigm.* Oxford: Springer.

International Safety Association of Oil and Gas Producers (2011). Process Safety - Recommended Practice on Key Performance Indicators. *Report No. 456.*

Liu, X., Saat, R., & Barkan, C. (2017). Freight-train derailment rates for railroad safety and risk analysis. *Accident Analysis and Prevention, 98*, 1-9.

Liyanage, J. P. & Kumar, U., (2003). Towards a value-based view on operations and maintenance performance management. *Journal of Quality in Maintenance Engineering,* 9, 4, 333-350.

Morrison, W. K., (February 2014). Select leading indicators to help measure safety. *Safety+Health.* <http://www.safetyandhealthmagazine.com/articles/9846-get-started-leading-indicators.

Molinski, D. (2015). How to Transport Oil More Safely. *The wall street journal*

Morris D. Z., (2012). Shipping crude oil by rail: Are the gains worth the risks? Retrieved from http://fortune.com/2015/07/02/crude-oil-shipping/

Nasir R. (2015). Oil shipping: rail versus pipeline? *CNBC.* Retrieved from http://www.cnbc.com/2015/02/17/oil-shipping-rail-versus-pipeline.html

Renne, J. (2013). *Transport beyond oil: Policy choices for a multimodal future.* Washington, DC: Island Press.

Tamminen, T. (2012). *Lives per gallon: The true cost of our oil addiction.* Washington, DC: Island Press.

Tippee, B. (2012). Producer cites benefits of rail transport of bitumen. *Oil and Gas Journal, 110*(7), 26-28.

Wardley-Smith, J. (2012). *The prevention of oil pollution.* New York: Springer Shop.